MY FIRST BOOK ABOUT THE SOLAR SYSTEM

All About the Planets for Kids

Dr. Melissa Caudle

My First Book About the Solar System

Publisher: Absolute Author Publishing House
Editor: Kathy Rabb Kittok
Cover Design: Rebeca @Rebeccacovers
Consultant: Jamie Alyson Dwyer
Child Consultant: John Axel Dwyer

ISBN Hardback: 978-1-64953-257-2
ISBN: Paperback: 978-1-64953-256-5
ISBN: eBook: 978-1-64953-258-9

Background pictures by Pixabay, Stock Nation, and NASA.

Planet, Comet, Astroid Graphics by NASA.

Dedication

To all the young space explorers who came before us and who will follow. Always keep dreaming because one day, the impossible becomes possible.

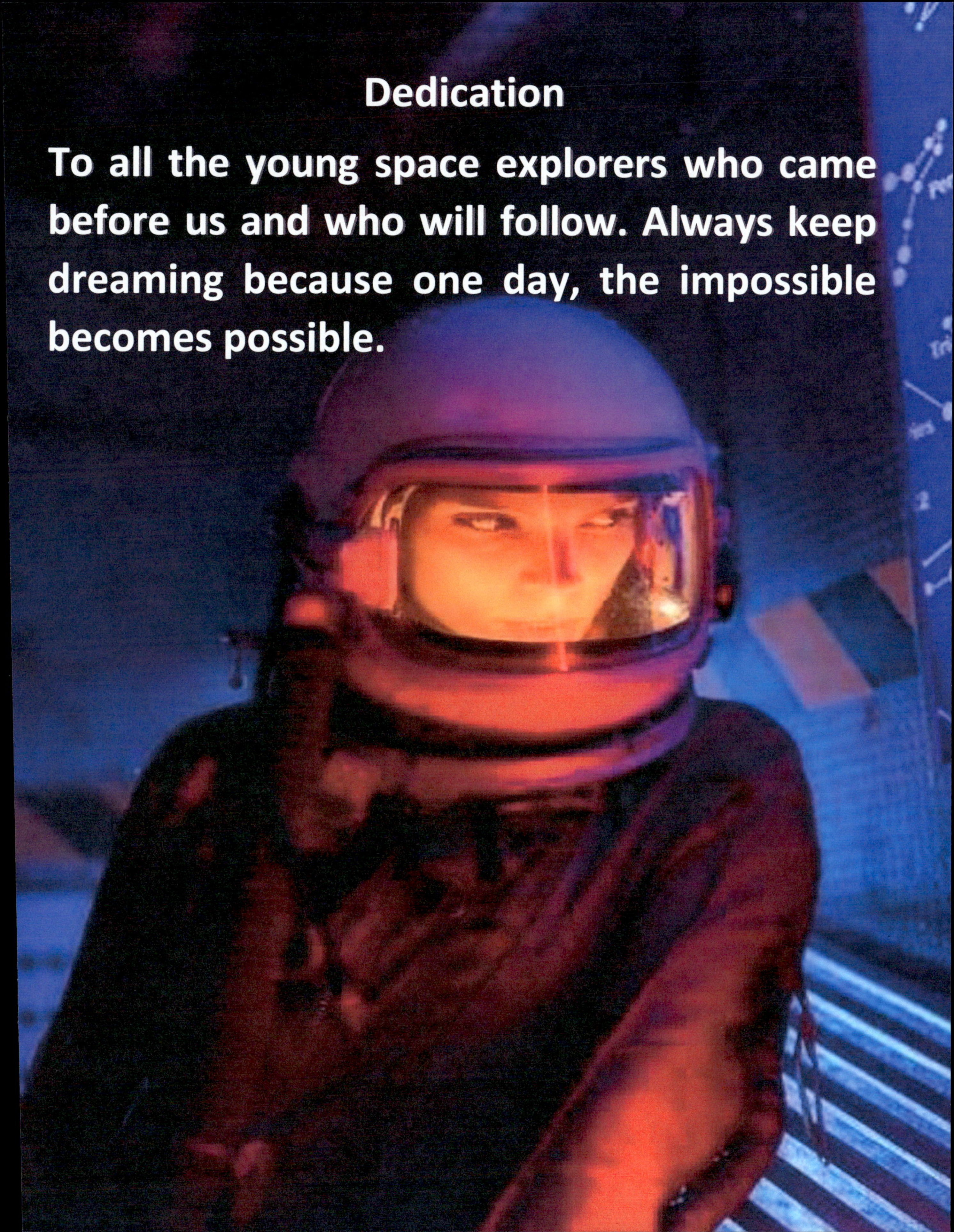

Can you name each planet, the Sun and the Moon?

OUR SOLAR SYSTEM

Did you know the solar system is made up of objects that spin, orbit, around the Sun? That's right, things like:

- Planets
- Moons
- Stars
- Rocks
- Asteroids
- Comets
- Dwarf planets, and
- Large amounts of dust

There are eight planets, six dwarf planets, the Sun, and the Moon in our solar system. The planets are categorized as terrestrial, gas giants, or dwarf.

The Big Two

Sun

Moon

Terrestrial Planets

Earth

Mercury

Venus

Mars

GAS GIANTS
Jupiter
Saturn
Uranus
Neptune

DWARF PLANETS

SUN

The Sun is not a planet but is one of more than 100 billion stars in our galaxy and is the largest object in our Solar System. It is about 70% hydrogen, 28% helium, and about 2% metals. One unique thing is that the Sun changes as it converts hydrogen to helium in its core. The surface of the Sun is called the photosphere, with a temperature of 5800K. The sunshine from the Sun is essential to Earth because, without it, we could not exist.

MOON

The Moon is the only natural satellite of Earth. Scientists believe that the Moon was created over 4.6 billion years ago. It is the fifth-largest moon in our Solar System and, after the Sun, is the brightest object in our night's sky. It is about 384,000km away from the Earth, which is about 239,000 miles.

EARTH

The Earth, one of the terrestrial planets, orbits the Sun. For a complete orbit, it takes about 365 days at a speed of 70,000 miles per hour. Its atmosphere is a mixture of gases which includes oxygen that we breathe. We need to take care of our planet by not polluting it, not overcutting our trees, and recycling old materials.

MERCURY

Mercury is the smallest and fastest planet in our Solar System. It takes Mercury 88 Earth days to circle around the Sun, and it is also the closest planet to the Sun. Surprisingly, Mercury is the most similar to Earth and is the second densest planet, and is about one-third the size of Earth.

VENUS

Venus is the second planet from the Sun, which is also the sixth-largest planet and can be seen with a telescope in our night's sky. Although we can see Venus, it is believed that clouds surround it. Surprisingly, Venus is the hottest planet.

MARS

Mars is the fourth planet from the Sun and has a very thin atmosphere. To date, it the most searched planet for life in our Solar System. Like the Moon, it has many craters and valleys, and like Earth, it also has many deserts and polar ice caps.

JUPITER

The fifth planet from the Sun is Jupiter. Jupiter holds the honor of being the largest planet in our Solar System. An interesting fact is that Jupiter was the first planet to take shape out of the remains of a solar nebula. Galileo Galilei conducted the first telescopic observations in 1609, and in 1610 Galileo also discovered the major moons of Jupiter, but of course not the smaller ones.

SATURN

The sixth planet from the Sun is Saturn and is most recognizable by the planetary rings that are around it. A surprising fact about Saturn is that it has more moons than Jupiter. It is about nine times larger than Earth.

URANUS

Located seven planets away from the Sun, sits Uranus. It is known as an ice-giant and is the third-largest planet in our Solar System. In 1781, Sir William Herschel announced Uranus' discovery and still holds the record of being the first planet discovered with the help of a telescope.

NEPTUNE

Neptune, first discovered in 1846, is the fourth largest and farthest planet in our Solar System. It also holds the record of having powerful winds, in fact, the strongest winds on all planets. Another first for Neptune is that it was the first planet to be discovered by mathematical predictions.

PLUTO

Pluto has a unique history in our Solar System. First, it is the largest known dwarf planet. Once thought to be the ninth planet for 75 years, things changed when astronomers discovered similar planets in 2006, which demoted Pluto to a dwarf planet. Other dwarf planets include Ceres, Haumea, Makemake, Eris, and Farout, which was discovered in 2018.

DWARF PLANETS

Besides Pluto, there are other dwarf planets in our solar system. They include Ceres, which is located in the Kuiper Belt between the orbits of Mars and Jupiter.

Located beyond the orbit of Neptune is the dwarf planet Haumea and was discovered in 2004. Of all the planets and dwarf planets, Haumea holds the honor of having a controversial shape in that it is more egged-shaped than spherical and has two moons called Hiiaka and Namaka.

Makemake, is the second furthest dwarf planet from the Sun. It was first discovered in 2005 by a team of astrologers at the Palamor Observatory and is the third-largest known dwarf planet.

Eris was discovered in 2005. An odd fact is that it was given the nickname of Xena and later give its name Eris. It is the second-largest dwarf planet and has one moon named Dysnomia. It takes 558 Earth years for Eris to orbit the Sun.

Farout was discovered on November 10, 2018, by the Japanese telescope, Subaru, is the farthest dwarf planet in our solar system. It takes over 1,000 years to orbit the Sun. It is the coldest dwarf planet. The real name is VG18, but astronomer Scott Shepard said "Farout," when he discovered it.

COMETS

A ridiculously small part of our Solar System includes comets which are small bodies and very icy. When a comet gets close to the Sun, the iciness begins to melt away, which is called outgassing and produces atmospheric trails called a coma or a tail. Comets can get as large as fifteen times the diameter of Earth. Comets are easily seen because they are bright. As of 2021, there are about 4600 known comets seen from Earth with the naked eye. Interestingly, the European Space Agency's Rosetta was the first unmanned probe to land a robotic spacecraft on a comet.

Five Parts to A Comet

- ✓ Nucleus
- ✓ Coma
- ✓ Hydrogen Cloud
- ✓ Dust Tail
- ✓ Ion Tail

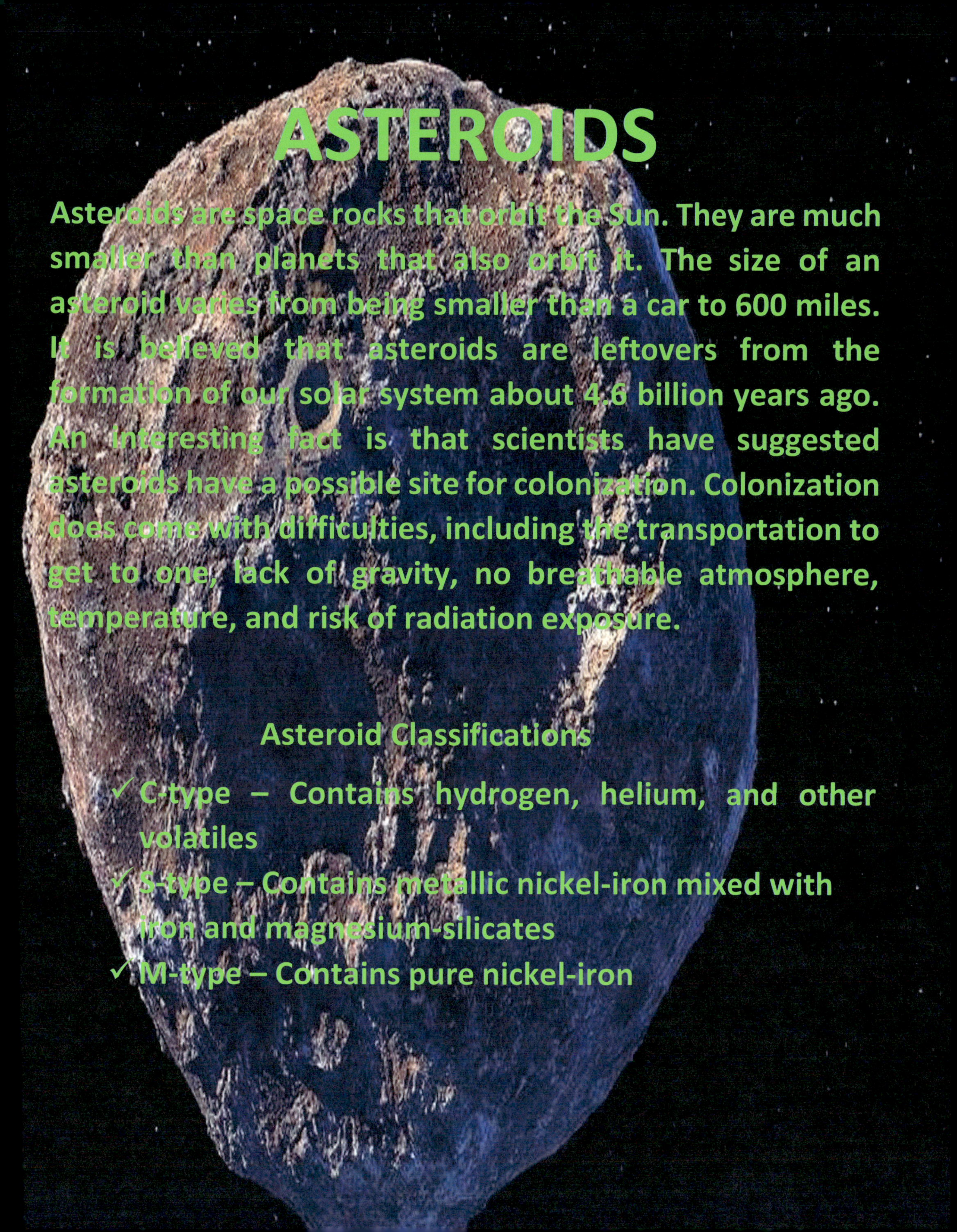

ASTEROIDS

Asteroids are space rocks that orbit the Sun. They are much smaller than planets that also orbit it. The size of an asteroid varies from being smaller than a car to 600 miles. It is believed that asteroids are leftovers from the formation of our solar system about 4.6 billion years ago. An interesting fact is that scientists have suggested asteroids have a possible site for colonization. Colonization does come with difficulties, including the transportation to get to one, lack of gravity, no breathable atmosphere, temperature, and risk of radiation exposure.

Asteroid Classifications

- ✓ C-type – Contains hydrogen, helium, and other volatiles
- ✓ S-type – Contains metallic nickel-iron mixed with iron and magnesium-silicates
- ✓ M-type – Contains pure nickel-iron

METEORS

Meteors are called fireballs or shooting stars. They range in size from small dust grains to small asteroids. When a meteor enters our atmosphere and hits the ground, it is called a meteorite. NASA tracks meteors and announces meteor showers, which can be seen with a naked eye.

FACTS AND MYSTERIES

1. Meaning the ground, the name for Earth, has been around for over 1,000 years.

2. Earth is the only planet that is not named after a Greek or Roman god.

3. Earth has an equatorial radius of 3,958 miles.

4. Earth's diameter is 7,917 miles.

5. Water covers 71% of Earth, and 29% is land.

6. The atmosphere of Earth has six layers: troposphere, stratosphere, mesosphere, thermosphere, exosphere, and ionosphere.

7. About 19 miles deep is the Earth's crust which is the outermost layer. The Earth's magnetosphere is a shield that protects us from solar and cosmic radiation.

8. The Earth rotates on its axis once every 24 hours.

9. Another name for the Moon is Luna by Romans and Artemis by the Greeks.

10. The Soviet spacecraft Luna2 was the first to visit the Moon in 1959.

11. On July 20, 1969, American astronaut Neil Armstrong was the first to walk on the Moon.

12. The first words spoken on the Moon were made by Neil Armstrong. He said, "That's one small step for man, and one giant leap for mankind."

13. Did you know that you can see the past by looking out into space? How? Deep space sends light toward Earth. When you look out into it, you are seeing light that was sent out from distant points many, many years ago. The longer it takes for the light to reach Earth, the older that light is. Cool, right?

14. Stars twinkle because light passes through the Earth's atmosphere, which has cold and hot air. When the light passes between the cold and hot air, it gives that twinkling effect because the light is coming from two different directions at once.

15. Not all planets will allow for exploration by landing on them. Why? A spacecraft can only land on a solid planet like Mercury, Venus, Earth, Mars, and Pluto. A spacecraft cannot land on the gas giant planets such as Jupiter, Saturn, Uranus, and Neptune, because they are huge balls of gas and liquid with no solid surface.

16. Mars is red because it contains lots of iron that has rusted away over millions of years.

17. There are numerous manmade satellites in space that can be seen from Earth. They blink very slowing.

18. To reach Pluto in an Apollo spacecraft like the one Neil Armstrong landed on the Moon, it would take 86 years for us to get there.

THE MILKY WAY

Our Solar System is in the Milky Way, which includes the Sun, a star, and all of the planets and dwarf planets you learned today.

OTHER BOOKS FOR CHLDREN BY DR. MELISSA CAUDLE

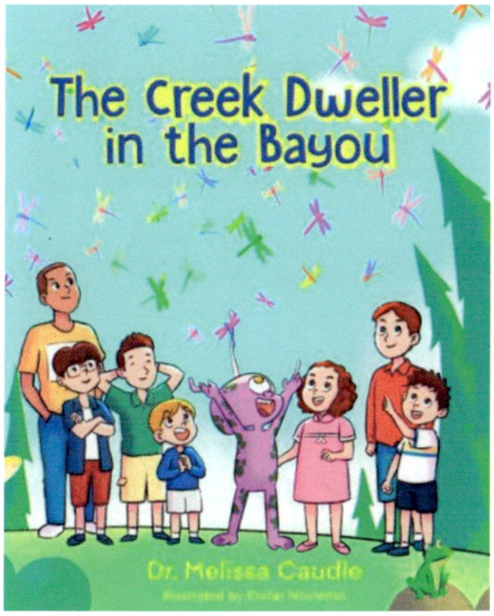

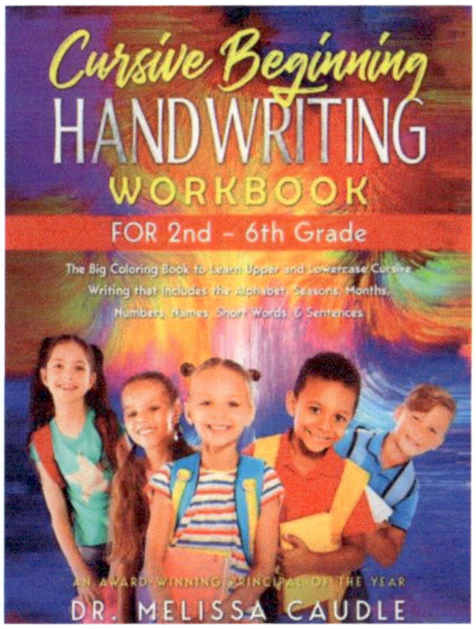

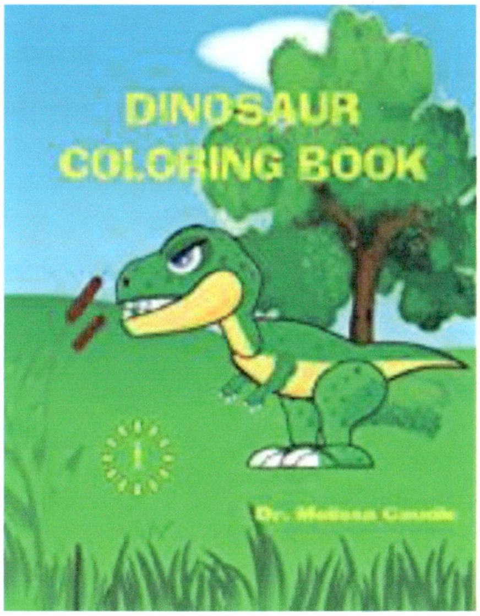

I LOVE UNICORNS
COLORING BOOK FOR GIRLS
2
Dr. Melissa Caudle

I LOVE UNICORNS
COLORING BOOK FOR GIRLS
1
Dr. Melissa Caudle

I LOVE UNICORNS
COLORING BOOK FOR GIRLS
3
DR. MELISSA CAUDLE

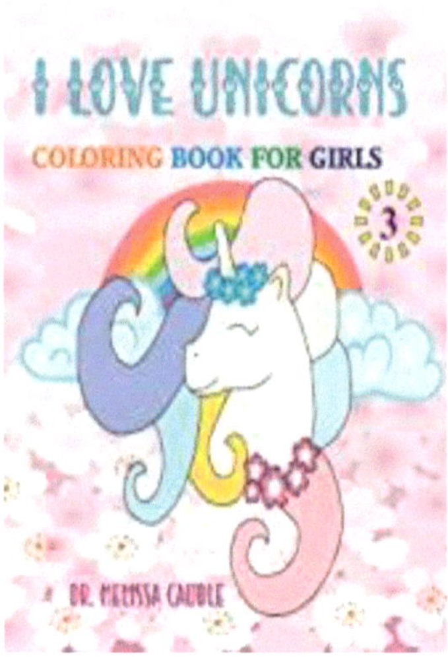
I LOVE UNICORNS
COLORING BOOK FOR GIRLS
3
DR. MELISSA CAUDLE

THE BIG COLORING
BOOK OF UNICORNS
DR. MELISSA CAUDLE

I LOVE BUTTERFLIES
COLORING BOOK FOR ALL AGES
Dr. Melissa Caudle
1

Made in United States
Troutdale, OR
11/03/2026

56845098R00019